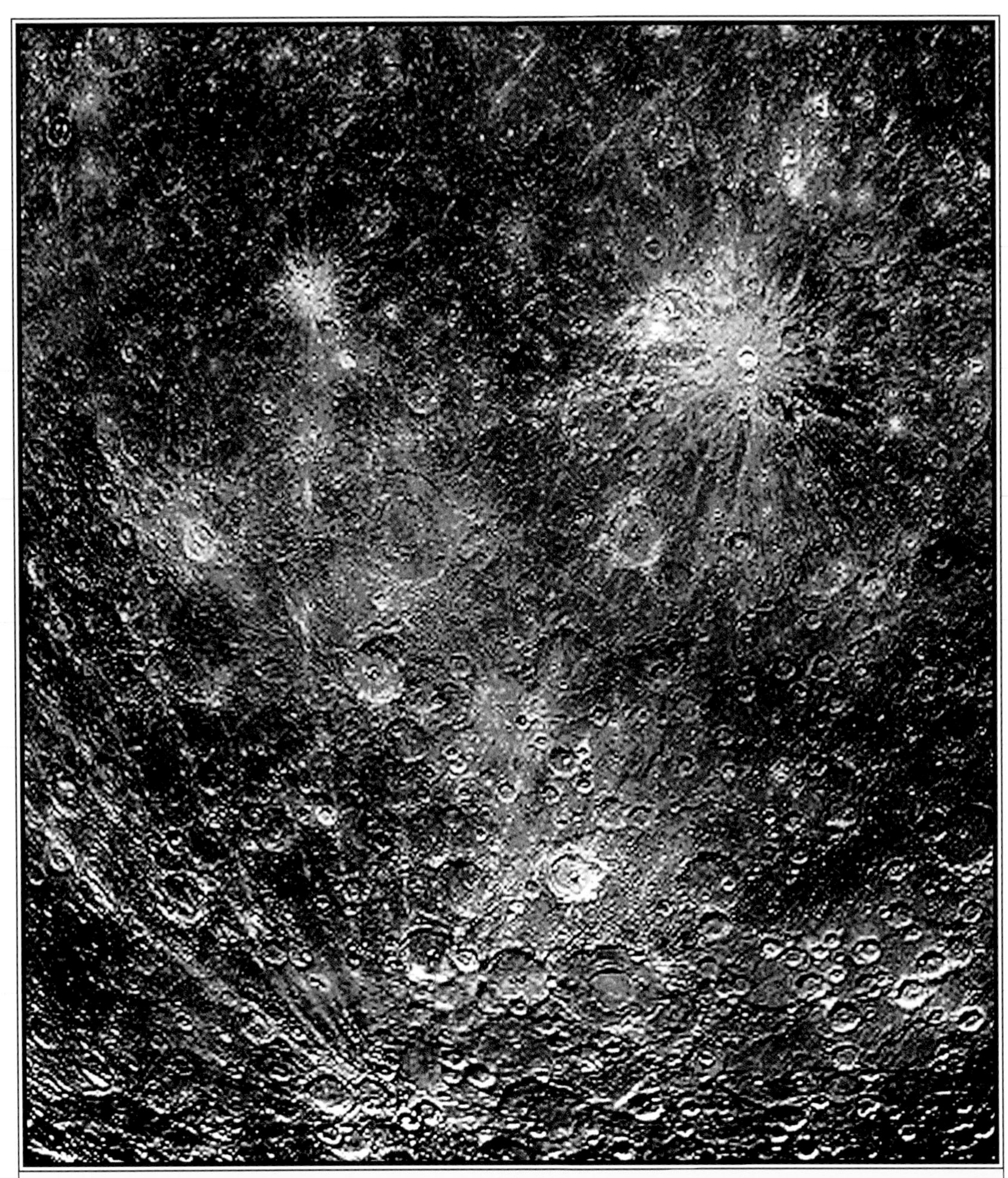

The pocked and cratered surface of Mercury

OUR SOLAR SYSTEM

Mercury

Steve Potts

A+
Smart Apple Media

C O P Y R I G H T

Published by Smart Apple Media
1980 Lookout Drive, North Mankato, MN 56003
Designed by Rita Marshall

Printed in the United States of America

Pictures by Bridgeman Art Library, Photo Researchers (Chris Butler/Science Photo Library, A. Gragera/Latin Stock/Science Photo Library, Ludek Pesek/Science Photo Library, Science Photo Library, U.S. Geological Survey/Science Photo Library, Detlev Van Ravenswaay/Science Photo Library), Tom Stack & Associates (Brian Parker, TSADO/NASA)

Library of Congress Cataloging-in-Publication Data
Potts, Steve. Mercury / by Steve Potts. p. cm. – (Our solar system)
Includes bibliographical references and index.

ISBN 1-58340-093-1

1. Mercury (Planet)–Juvenile literature. [1. Mercury (Planet)] I. Title.
QB611 .P68 2001 523.41–dc21 2001020123

First Edition 9 8 7 6 5 4 3 2 1

CONTENTS

Mercury

Nearest the Sun

For the ancient Romans, Mercury was the god of commerce and travel, the winged messenger. Like the god Mercury, who moved quickly across the sky, the planet Mercury also moves quickly. In fact, of all the planets, Mercury is the one that moves the fastest around the Sun. ✹ Mercury is the second smallest planet. It is less than half the size of Earth. Until space exploration began in the 1960s, astronomers

Mercury's day is 58.65 Earth days long; its year is 87.97 Earth days long.

The Roman god Mercury was known for his speed

(scientists who study planets) knew little about Mercury. Today, much more information is available about the Sun's nearest neighbor.

About Mercury

Mercury is close enough to Earth that astronomers such as Nicolaus Copernicus could see it in the 1500s without any equipment. Despite the many years that scientists have been studying this planet, there are still many things about Mercury that remain a mystery. Scientists believe that the core of the planet, its center, is made mostly of iron and nickel.

This core, the densest of any in the solar system, accounts for about four-fifths of Mercury's size. It is not known for sure what covers this solid metal center. Mercury's dense core

Scientists think Mercury has a center made of iron

From left to right: the Sun, Mercury, Venus, Earth, Mars

may also be responsible for creating the planet's magnetic field, which acts like a shield against the Sun's harsh wind. Unlike Earth, Mercury has a very thin **atmosphere**. That means there is little to protect Mercury's surface from rapid temperature changes. On the side of Mercury that faces the Sun, surface temperatures can reach 872 °F (467 °C). On the side of Mercury that is turned away from the Sun, temperatures drop to –298 °F (–183 °C). Astronomers believe that Mercury's thin atmosphere is 42 percent oxygen. This is a higher level

Mercury is only about 36 million miles (58 million km) from the Sun.

than on Earth. Scientists don't know what produces this high

oxygen level.

The side of Mercury facing the Sun gets very hot

Mercury's Surface

Mercury's surface is similar to that of Earth's moon. About four billion years ago, pieces of rock from outer space hit Mercury. These falling rocks, called meteorites, left Mercury with many large craters. They also caused volcanic activity that spread lava all over the planet's surface. Meteorites, not wind or water, shaped Mercury's landscape. Since the meteorites hit Mercury, its crater-covered surface has not changed much.

Mercury's surface also includes a collection of surface ridges. Billions of years ago, Mercury was much hotter and

rotated much faster on its **axis**. As its rotation began to slow down, ridges formed on its surface that looked like wrinkles. When temperatures on Mercury began to cool, the surface

Long ago, Mercury was bombarded by meteorites

began to shrink. These ridges were pushed up, forming the giant ridges that cover many parts of Mercury's surface today.

The origin of the wrinkle ridges on Mercury's surface is arguable, however. Some scientists claim that **tectonic activity** caused the wrinkles, and others think that they are the result of thousands of volcanic flows. A unique area of Mercury looks like a patchwork of complex blocks. The planet's top layer of rock was shattered into pieces by a mas-

The Caloris Basin, Mercury's largest surface crater, is about the size of Texas.

The Caloris Basin and other meteorite craters

sive shock wave. The wave was produced when a giant mete-

orite struck the planet, creating the Caloris Basin.

Exploring Mercury

Although Mercury is one of the planets that is closest to Earth, it has had few visits from spacecraft. In 1974 and 1975, a space probe called *Mariner 10* flew by Mercury several times. It got within 470 miles (756 km) of Mercury and was able to photograph about 40 percent of the planet. ✹ Probes are small spacecraft that are sent into space by rockets or carried on a space shuttle. These probes carry complex computer and photographic equipment. The pictures they take are sent

Probes sent to Mercury must withstand great heat

back to Earth by **radio waves**. Much of what we know about Mercury came from information sent back by *Mariner 10.*

The U.S. plans to launch *Messenger*, an advanced space probe, to Mercury in 2004. By 2009, *Messenger* will orbit Mercury and begin sending information back to scientists on Earth so they can learn more about this mysterious planet.

Mercury has no life, but its surface may have water that contains oxygen.

Space probes took a close look at Mercury in the 1970s

Mercury is the innermost planet in our vast solar system

INFORMATION

Index

Words to Know

atmosphere—the nearly invisible layer of gases that surrounds a planet

axis—a non-moving, imaginary line that an object rotates around

radio waves—energy that travels at the speed of light (186,000 miles/second or 300,000 km/sec) from a transmitting antenna to a receiving antenna to form a message

tectonic activity—the movement of a planet's rocky crust to create basins, mountains, and other formations

Read More

Bond, Peter. *DK Guide to Space.* New York: DK Publishing, 1999.

Couper, Heather, and Nigel Henbest. *DK Space Encyclopedia.* New York: DK Publishing, 1999.

Furniss, Tim. *Atlas of Space Exploration.* Milwaukee, Wisc.: Gareth Stevens Publishing, 2000.

Internet Sites

Astronomy.com
http://www.astronomy.com/home.asp

NASA: Just for Kids
http://www.nasa.gov/kids.html

The Nine Planets
http://seds.lpl.arizona.edu/nineplanets/nineplanets

Windows to the Universe
http://windows.engin.umich.edu